Mariana Borges Barreto Bezerra
Ana Paula Pereira da Silva

SIDNONES: CHEMICAL AND ANTITUMOR PROPERTIES

Mariana Borges Barreto Bezerra
Ana Paula Pereira da Silva

SIDNONES: CHEMICAL AND ANTITUMOR PROPERTIES

A narrative review

ScienciaScripts

Cover image: www.ingimage.com

This book is a translation from the original published under ISBN 978-620-6-75943-0.

Publisher:
Sciencia Scripts
is a trademark of
Dodo Books Indian Ocean Ltd. and OmniScriptum S.R.L publishing group

120 High Road, East Finchley, London, N2 9ED, United Kingdom
Str. Armeneasca 28/1, office 1, Chisinau MD-2012, Republic of Moldova, Europe
Printed at: see last page
ISBN: 978-620-7-97266-1

DEDICATORY

I dedicate this work to my grandmother Irinéa.
For allowing and supporting all the women in our family to study. And for always teaching us the importance of being independent. This is another diploma that's all yours, Grandma!

ACKNOWLEDGMENTS

To God, who has always been the support of my life, and to Our Lady, for her enormous intercession in all my plans.

To my mother Lídice, who always supported all my dreams and taught me never to give up. You are my greatest example as a woman, daughter, mother and wife. Thank you for believing, Mom! We did it!

To my father Lupércio, who is simply the best father in the world. With you, I learned to put love into everything I do and always do it well. Thank you for all your care and advice.

To my sister Ana Clara, for being my confidant, best friend and the greatest gift I've ever received. Our complicity is unique. Thank you for wiping away my tears and always cheering me up. I love you to the moon, back and forth.

To my grandparents, for all the love they have always shown me. To my eternal grandmother Irinéa, who taught me love and everything that is good in this world. And to my grandfather Adelson, for teaching me the value of an honest life and for all the trips to the bus stop just so I wouldn't be alone.

To my fiancé Rodrigo, for all the companionship on this journey we've shared. Thank you for being patient, understanding and for encouraging me every day to strive for more.

To my friend Paloma, who has never abandoned me since the first day of class. In good times, you were happy with me and in sad times, you consoled me. Thank you for calling me to every event, even though you knew I wouldn't go, and for always looking out for me. Thanks also to my friend Juliana, who always shared her knowledge with me. I'll never forget when you took my hand so that I could pass Organic 2!!!.

To my advisor Prof. Dr. Ana Paula Pereira da Silva, for all the knowledge shared during this journey; and, above all, for all her understanding and patience, doing everything possible for this work to be completed.

To my lab friend William. Thank you for all your help in this world of biochemistry. Your partnership was very important for my academic and personal growth.

To the Federal Rural University of Rio de Janeiro - Seropédica Campus, for everything it has given me over these five years. Its wonderful structure, excellent teachers, committed staff and unique beauty have all contributed to my professional development.

Finally, I would like to thank all the people who directly or indirectly encouraged me and helped me to complete this scientific work.

SUMMARY

Bezerra, Mariana Borges Barreto. Sidnones: a review of their chemical and antitumor properties. Seropédica, 2021. Course Conclusion Paper (Industrial Chemistry) - Chemistry Institute, Federal Rural University of Rio de Janeiro.

Mesoionic compounds are synthetic substances that belong to the class of heterocyclic betaines. Among this class, the sidnones stand out. The *sydnones* were first prepared by Earl and Mackney at the University of Sydney in 1935 and this gave rise to their name: *sydnone.*They have structures, chemical properties, synthetic utility and biological activities that are considered relevant. One of these characteristics is their possible anti-cancer effect. Cancer is one of the main causes of death and every year more people are affected by the disease. Current treatments have low efficacy and many side effects, which is why there is a great desire for new, more effective drugs to treat this disease. For this reason, the aim of this work was to provide a narrative review to analyze the potential of siddnones for future use in cancer therapy, as well as detailing their chemical and unique properties.

KEYWORDS: Sidnones, antitumor, mesoionic compounds.

ABSTRACT

Bezerra, Mariana Borges Barreto. Sidnones: a review of their chemical and antitumor properties. Seropédica, 2021. Course Conclusion Paper (Industrial Chemistry) - Chemistry Institute, Federal Rural University of Rio de Janeiro.

Mesoionic compounds are synthetic substances that are part of the heterocyclic betaine class. Among this class, the sydnones stand out. Sydnones were first prepared by Earl and Mackney at the University of Sydney in 1935 and this gave rise to their name: sydnone. They have structures, chemical properties, synthetic utility and biological activities considered relevant. One of these features is its possible anti-cancer effect. Cancer is a leading cause of death and every year more people are aggravated by the disease. Current treatments have low efficacy and several side effects, so there is a great desire for new drugs that are more effective in the treatment of this disease. For this reason, the aim of this paper was to provide a narrative review to analyze the potential of sidnones for future use in cancer therapy, as well as to detail their unique chemical properties.

KEYWORDS: Sydnones, antitumor, mesoionic compounds.

LIST OF ABBREVIATIONS

A498	Human kidney cancer cell line
A549	Human lung cancer cell line
BT 474	Breast cancer cell line
CA4P	Combretastatin A4
DAPI	(4',6-diamidino-2-phenylindol)
DNA	Deoxyribonucleic acid
HeLa	Human cervical carcinoma cell line
IUPAC	International Union of Pure and Applied Chemistry
Jurkat	Human lymphocyte leukemia cell line
MTT	((3-(4,5-dimethyl-2-thiazolyl)-2,5-diphenyltetrazoliobromide).
SW1222	Adenocarcinoma of the human rectal cervix
SYD-1	[N-(4-chloro-3-nitrophenyl)-sidnone]
UV	Ultraviolet spectrum

SUMMARY

1INTRODUCTION

he World Health Organization (WHO) estimates that by 2030 there will be 27 million new cases of cancer, with 17 million deaths from the disease. According to the WHO, developing countries account for more than 70% of all cancer deaths worldwide. In Brazil, cancer is also one of the main health problems and for each year of the 2020/2022 triennium, it is estimated that 625,000 new cases of cancer will be registered. In 2019, approximately 230,000 cancer deaths were recorded in Brazil (INCA, 2019). Both cancer incidence and mortality have been increasing worldwide. Factors such as population growth, ageing and socioeconomic development may be related to this increase (BRAY et al., 2018). Unfortunately, truly effective treatments for this disease are not yet available.

With regard to chemotherapy, its side effects and low efficacy mean that the development of more efficient antitumor agents that are less toxic to normal cells is extremely important and urgent (SUNG et al., 2021). In this context, one class of compounds that has been shown to be relevant due to their antitumor action is mesoionic compounds (GOZZI et al., 2015), especially sidnones. Sidnones are molecules with unique chemical properties that have broad potential in medicinal chemistry. Although still in the development phase, these compounds have shown significant results in terms of their anti-tumor properties. Therefore, this paper aims to present the main characteristics and chemical properties of sidnones and the most relevant data available in the literature on their antitumor activity. In this way, scientific progress in this area can be identified and the future use of sidnones in cancer treatment evaluated.

2JUSTIFICATION

In the search for new substances with anticancer activity, sidnones stand out. Sidnones are compounds belonging to the mesoionic class and have shown significant anticancer activity *in* both *in vitro* and *in vivo* tests.

Therefore, a survey and analysis of the main characteristics of sidnones and their effect against cancer in research carried out over the last decade could be a useful tool in evaluating the potential of these compounds as future antitumor agents.

3OBJECTIVES

3.1 General objective

The aim of this narrative review was to analyze the potential of sidnones for future use in cancer therapy.

3.2 Specific objectives

The specific objectives of this work were:

1. Present the main characteristics and chemical properties of sidnones;
2. Present the main results in the literature on the antitumor activity of different sidnones.

4 MATERIAL AND METHODS

4.1 Bibliographic search

To carry out this work, a bibliographic survey was carried out on the PUBMED electronic database. The search strategy was devised using the descriptors "SYDNONE" and "CANCER" and the Boolean operator "AND". The searches were carried out during the first and second semester of 2021 and included original articles, review articles, clinical trials and case reports, published in English and Portuguese, with a publication period. For data on cancer and the chemical properties of sidnones, no restrictions were used for the period of publication. For data on the use of sidnones as antitumor agents, a filter was used for publications from the last 10 years.

5 RESULTS AND DISCUSSION

5.1 What is cancer

Cancer is a term that characterizes more than 100 types of malignant diseases. The common characteristic among these diseases is the uncontrolled growth of cells that are capable of invading organs and tissues (INCA, 2021).

The term cancer refers to the uncontrolled or malignant growth of cells that have the ability to invade tissues and organs. Because of this rapid division, tumor cells tend to be very aggressive and lead to the formation of tumors that can spread to other regions of the body (a process called metastasis) (INCA, 2021).

The types of cancer vary according to the types of cells in the body. When they are in epithelial tissues, such as skin or mucous membranes, they are identified as carcinomas. If they are in connective tissues, such as bones, cartilage or muscle, they are called sarcomas (Figure 1) (INCA, 2021).

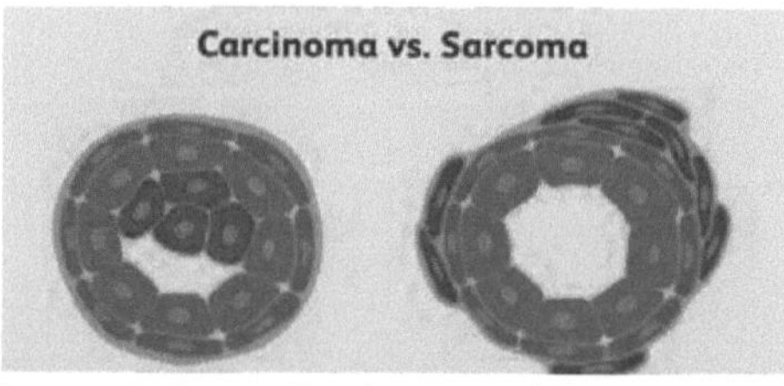

Figure 1. Illustrative difference between Carcinoma and Sarcoma.
Source: https://www.verywellhealth.com/angiosarcoma-of-the-breast-430619

The onset of cancer can be attributed to various causes, internal and/or external to the body. Of all cancer cases, 10 to 20% are associated wh internal causes, being genetic factors or linked to the body's ability to defend itself against external aggressions. On the other hand, between 80% and 90% of cancer cases are associated with environmental risk factors called carcinogens, which are external factors (SOUSA-PEREIRA, 2019).

Cancer can be treated with a combination of surgery, radiotherapy, chemotherapy or transplants (INCA, 2021). Despite advances in the understanding of bio-organic procedures linked to carcinogenic potential, there are still huge challenges for effective cancer treatment, since conventional chemotherapeutic agents can be not only ineffective, but also generally toxic to the patient's body.

It's important to note that not every tumor is called cancer. A tumor is an increase in volume observed in any part of the body.

When this tumor is caused by an increase in the number of cells, it is called a neoplasm. It can be benign or malignant (Figure 2) (INCA, 2021).

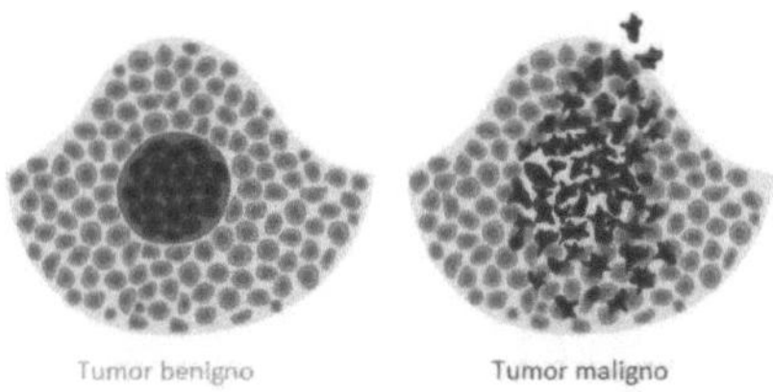

Figure 2: Representation of the organization of cells in benign and malignant tumors Source: https://www.biologianet.com/doencas/neoplasia

Benign neoplasms are organized and generally slow-growing, and have precise boundaries. They don't even invade neighboring tissues or develop metastases (INCA, 2021). Although they don't cause metastases, they can compress neighboring organs and tissues. Lipoma and myoma are examples of benign tumors.

Malignant neoplasms, or cancers, have greater autonomy and can invade adjacent tissues and cause metastases. They can be resistant to treatment and can also lead to death (ROBBINS, 2004).

5.2 Mesoionic compounds

5.2.1 General Structure

Mesoionic compounds are synthetic substances that are part of the heterocyclic betaine class, which in turn are flat molecules composed of five atoms, with at least one side chain, presenting dipole moments and the α atom in the same plane as the ring (Figure 3) (SILVA *et al.*, 2018).

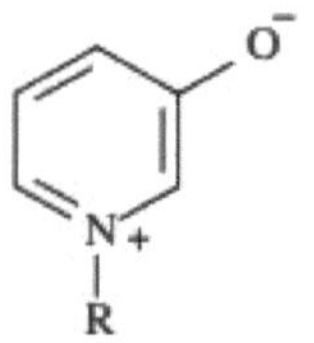

Figure 3: Chemical structure of betaines
Source: KIER &. ROCHE (1967).

The term "mesoionic" (mesomeric + ionic) was coined by Baker *et al.* in 1949 to indicate the inherent ionic character of these heterocyclics (BADAMI, 2006). The first mesoionic compound was synthesized by Fisher and Besthorn over a hundred years ago (OLLIS *et al.*, 1985). Represented by the bicyclic structure, it was called "dehydrotizone" at the time (Figure 4).

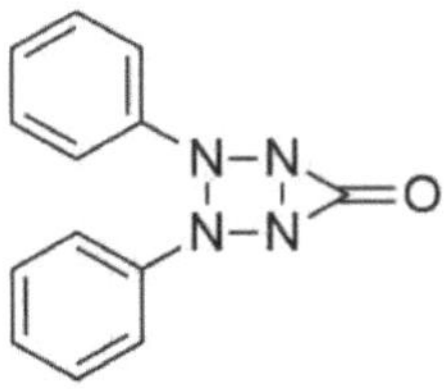

Figure 4 - Representation of the chemical structure of the first mesoionic compound synthesized: dehydrotizone. Source: OLLIS *et al.* (1985).

In 1955, Baker and Ollis proposed that if a compound had a five- or six-membered heterocyclic structure that could not be adequately represented by a polar or covalent structure, it could be called mesoionic. In addition, it must have a π-electron sextet attached to all the atoms that make up the ring. Together, a positive charge must exist on the heterocyclic and an atom or group of atoms attached to the ring with opposite charges (BRANDT *et al.*, 2014).

Over the years, this concept of mesoionics has changed. In 1957, Baker and Ollis suggested the term "mesoionic" to describe heterocyclic compounds with both mesomeric and ionic characteristics (BAKER *et al.*, 1957). In 1976, Ollis and Ramsden decided to change the concept applied to these compounds in order to limit the use of the term "mesoionic" to 5-membered heterocyclic compounds only (OLLIS *et al.*, 1976).

The IUPAC (International Union of Pure and Applied Chemistry) defines mesoionic compounds as consisting of a 5-membered heterocyclic ring that cannot be correctly represented by a covalent structure or just a polar structure, thus being described by multiple structures.

This ring has a sextet of π electrons, supporting positive charges counterbalanced by negative charges on the exocyclic atom, a, b, c, d, e, f commonly being carbon, sulphur, nitrogen, oxygen or selenium (Figure 5) (OLLIS & RAMSDEN, 1976; NEWTON & RAMSDEN, 1982).

The mesoionic ring has an aromatic character and forms a plane together with the alpha atom of the heterocyclic ring (ECHEVARRIA *et* al., 1992; KAI-CHUNG *et* al., 1993).

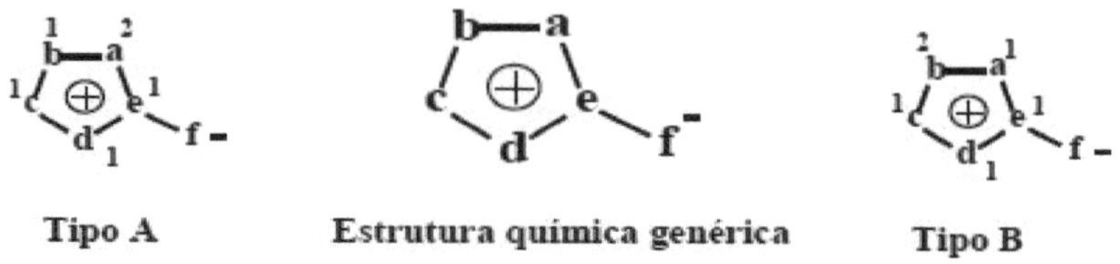

Figure 5: Generic chemical structure of mesoionic compounds and representative structures for type A and B compounds. Source: SOUSA (2019).

The combinations of structural characteristics described above promote interactions with biomolecules such as DNA and proteins and, in this way, mesoionic compounds have been the subject of study in pharmaceutical applications, since their unique structures favor a great diversity of biological activities such as anti-inflammatory, antibiotic, antifungal and antitumor (SILVA, 2018).

5.2.2 Sidnonas

Sydnones are the most studied compounds in the mesoionic family. This class of compounds was first prepared by Earl and Mackney at the University of Sydney in 1935 and hence the origin of its name: *sydnone*

Researchers at the Romanian Academy of Organic Chemistry

published their first paper in 1965 (SUCIU *et al.*, 1965) and since then numerous articles have been published on their synthesis reactions, properties and applications.

Sidnones have structures, chemical properties, synthetic utility and biological activities that are considered to be quite interesting. Since they were first prepared, sidnones have attracted the attention of medicinal chemists and pharmacologists to investigate their biological applications.

Because it has a different chemical structure, a variety of biomolecules such as DNA and enzymes can be switched on and off by it. A huge number of pharmaceutical properties have been attributed to sidnones, such as: antimicrobial, anti-inflammatory, anticancer, antioxidant and antidiabetic (CHANDRASEKAR, 2012).

Their aromaticity, as well as the 1,3-cycle addition reactions they result in, make these compounds a valuable tool for the synthesis of new heterocycles with complex structures (Figure 6) (ALBOTA *et al.*, 2017).

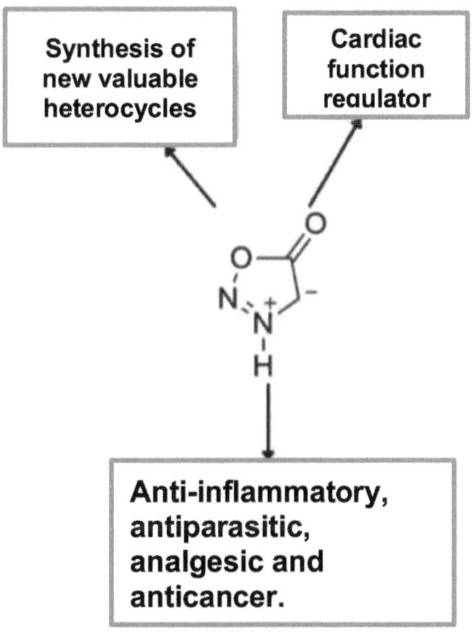

Figure 6. Applications of sidnones. Source: Adapted from ALBOTA *et al.*, (2017).

5.2.2.1 Structure

The formation suggested by the initial researchers was that of a three- to four-membered ring product (Figure 7), from the action of acetic anhydride on N- nitrosophenylglycine (BAKER AND OLLIS,1946).

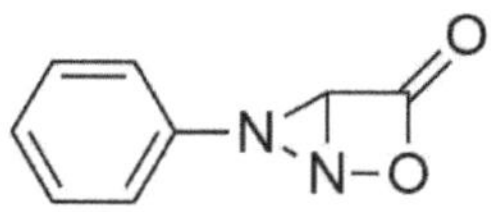

Figure 7. First structure proposed by Earl and Mackney in 1935.
Source: (ABDUALKADER *et al.*, 2017).

Later, two factors were discussed that revealed why this bicyclic system proposed by Earl was unlikely: at first, it is uncertain whether a fused ring system is the result of a simple intramolecular rearrangement. It is also a highly unstable structure due to the existence of a β-propiolactone group (Figure 8) (ABDUALKADER *et al.*, 2017).

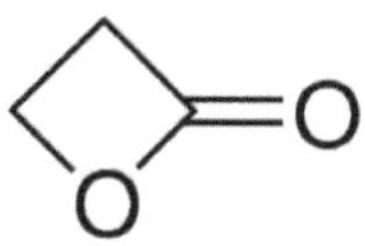

Figure 8. Structure of β-propiolactone.
Source: https://www.sigmaaldrich.com/content/dam/sigma-aldrich/structure3/177/mfcd00005169.eps/_jcr_content/renditions/mfcd00005169-medium.png

Previously, the bridging bond was omitted by Baker and Ollis (1946).

They proposed a partially aromatic five-membered ring (A and B) that was a hybrid of many zwitterionic forms (Figure 9) (BAKER and OLLIS, 1946).

AB

Figure 9. Partially aromatic rings proposed by Baker et al.
Adapted: ABDUALKADER *et al.*, (2017).

Secondly, acid hydrolysis decomposes the sidnone into hydrazine, carboxylic acids and carbon dioxide, while hot aqueous sodium hydroxide can cause the sidnone to return to the initial N-nitroso (EADE AND EARL, 1946).

In addition, other researchers have shown that acetic anhydride can convert N-nitroso-N-phenylalanine into the optically inactive N-phenyl-C-methylsidnone (Figure 10).

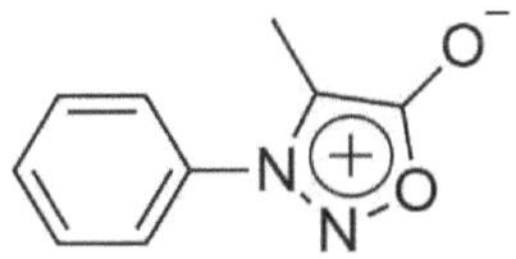

Figure 10. N-phenyl-C-methylsidnone. Source: ABDUALKADER *et al.*, (2017).

It is significant to say that the loss of activity entails racemization or a change in the hybridization of C4 from a chiral sp3 state to a chiral sp2 state (KENNER AND MACKAY, 1946).

The oxygen atom attached to C5 in the proper enolate format for rapid formation of a Bromine mono derivative (Figure 11) in glacial acetic acid and bromine (KENNER AND MACKAY,1946).

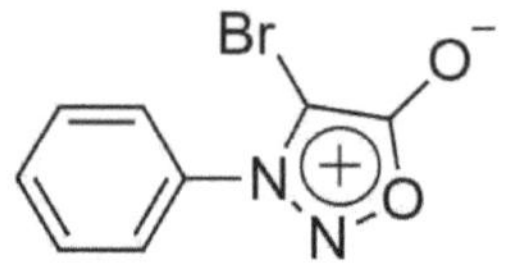

Figure 11. Bromine mono-derivative made in acetic acid and bromine. Source: ABDUALKADER *et al.*, (2017).

Baker and Ollis reviewed all the proposed structures for sidnones and, to end the debate about their first publication, suggested a more concise description. They admitted that the previous simple covalent structures did not characterize sidnones so well. This is because the sidnone molecule should be seen as a hybrid of bipolar and tetrapolar forms whose contribution to the hybrid ring is not the same. As a result, sidnones are described as mesoionic compounds (ABDUALKADER *et al.,* 2017).

The sidnone ring has all the requirements to develop aromatic properties. The exocyclic oxygen atom provides an electron to the ring to complete the π-electron sextet (BAKER AND OLLIS, 1946). With this, the pair defined the following structure (Figure 12) as representative of the sidnone class:

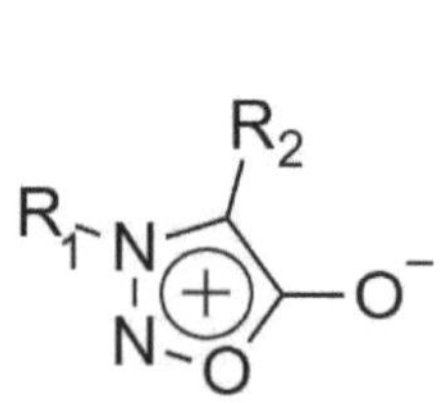

Figure 12. Structure defined by Baker and Ollis as the structure of sidnone. Source: ABDUALKADER *et al.*, (2017).

5.2.2.1.1 Flatness

The planarity of the sidnone ring is also notable for its special structure. This planarity can be verified by X-ray analysis on a large number of

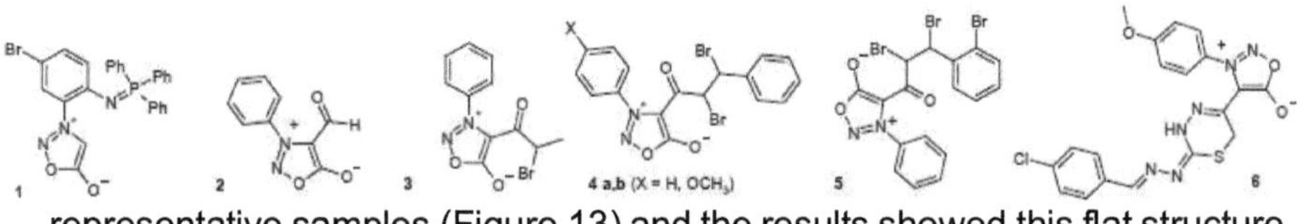

representative samples (Figure 13) and the results showed this flat structure for the heterocyclic ring. It is worth noting that, according to the analysis carried out, compounds with a carbonyl substituent (2,4-6) show a smaller deviation but also show this conjugation with the ring (ALBOTA *et al.*, 2016).

Figure 13. Different sidnones analyzed in the X-ray spectrum by ultraviolet spectroscopy (UV). Source: ALBOTA *et al.,* (2016)

The UV-Vis spectra also provide further references on the conjugation of the sidnone ring with the external double bonds, revealing the importance of steric factors. Stewart, Kier and Roche reviewed the ultraviolet spectra of sidnones. In summary, the absorption maxima in the 290-340 nm range prove the presence of the aromatic ring in siddnones (KIER AND ROCHE,1967).

A number of factors can significantly affect the UV spectra of sidnones: firstly, conjugation where an aromatic system substituted at C4 of the ring has a stronger bathochromic effect. This effect, also called red-shift, promotes an increase in lmax due to the effect of the solvent or a substituent group. So, for example, 3-methyl-4- phenylsidnone will have a UV maximum at 317 nm. Similarly, 4-acetylated sidnone absorbs at a longer wavelength, as 4-acetyl-3-phenylsidnone and 3-phenylsidone absorb at 324 and 310 nm (ABDUALKADER *et al.,* 2017).

Steric elements also interfere with lmax values. They can decrease conjugation by altering the planarity of the molecule. The UV maxima of 3-(2, 6-methylphenyl) sidnone were found at 255 nm, lower than those found in 3-alkyl sidnone which has little conjugation (ABDUALKADER *et al.*,2017).

Finally, the electrostatic interaction in the bis-sidnone system (Figure 14) makes the coplanarity system more rigid and therefore the UV absorption wavelength is unusually high as in: whose maximum absorption was at 350nm (ABDUALKADER *et al.*, 2017).

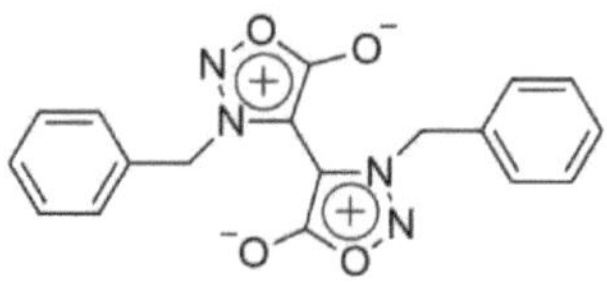

Figure 14. Bis-sidnone with a high lmax value. Source: ABDUALKADER *et al.*, (2017).

5.2.2.2 Aromaticity of sidnones

The aromaticity of sidnone is supported by its flat structure, delocalized charges and considerable resonance energy (GROSSIE *et al.*, 2013).

One of the main concepts in organic chemistry is aromaticity. It isbased on chemical stability, resonance energy and the possibility of undergoing substitution reactions (CARDOSO, 2021).

In simple terms, aromaticity has to do with the stability of benzene and its derivatives, as well as other unsaturated compounds. It determines that the structure, stability and reactivity of many molecules have properties defined by the cyclic delocalization of electrons, which end up being expressed as greater stabilization of aromatic compounds and destabilization of non-aromatic compounds (WOLKE, 2021).

One of the characteristics of aromatic compounds is their affinity for substitution. Compounds derived from benzene, aromatic hydrocarbons and their conjugated heterocyclic compounds are characterized by substitution reactions. But it is significant to say that compounds such as azulene and others, which perform substitution reactions, do not have other aromatic properties (BRAZ, 1992).

Hückell (1979), based on molecular orbital calculations, suggested that aromatic ring systems should contain an amount of electrons equal to (4n+2), where "n" is an integer and positive. As a result, heterocyclic and homocyclic compounds containing 2,6,10... electrons would be aromatic (BRAZ, 1992).

Looking at the structure of mesoionic compounds, they are considered aromatic, since their structures can be described as cyclic systems with p orbitals containing six electrons .

Duffin and Kendal (1956) studied electrophilic substitution reactions such as bromination and nitration in derivatives of the mesoionic system 1,2,3-thiadiazolium-4-olate (Figure 15) (BRAZ, 1992):

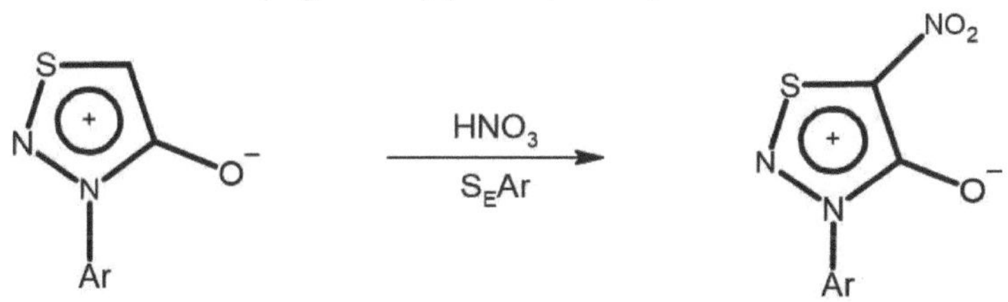

Figure 15. Mesoionic 1,2,3-thiadiazolium-4-olate system. Adapted from BRAZ (1992).

5.2.2.3 Summary

Eighty years ago, a sidnone was prepared by Earl and Mackney by the cyclodehydration of acetic anhydride in the N-nitroso derivatives of amino acids. They wrote that the dissolution of N-nitroso-N-arylglycine in an excess of acetic anhydride at room temperature resulted, after 24 h, in a stable, crystalline, nitroso-free heterocyclic product that was later named sidnone (APPLEGATE, 1988).

The N-nitroso intermediate was prepared by the conventional nitrosation of the amino group of N-phenylglycine by nitrous acid generated from the reaction of sodium nitrite with hydrochloric acid (BAKER AND OLLIS,1950).

Baker later deduced the cyclization mechanism of the N-nitroso starting material by the loss of a water molecule, which involves four steps, as shown in (Figure 16) (ABDUALKADER *et al.*, 2017).

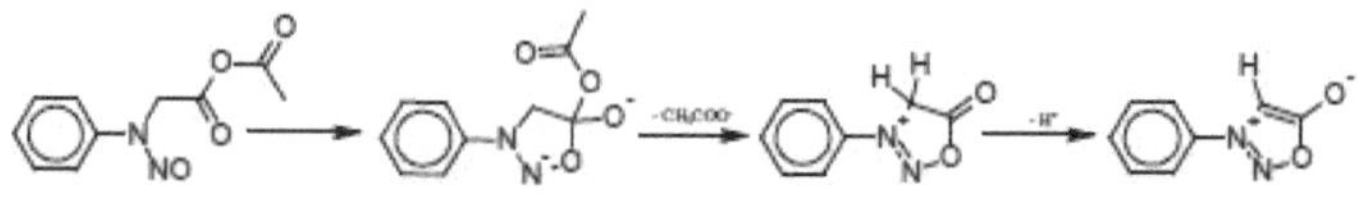

I II III IV

Figure 16. Mechanism of cyclization from N-nitroso. Adapted from ABDUALKADER *et al.*, (2017).

First, a mixed anhydride intermediate (I) will be formed from the effect of acetic anhydride on free nitrous acid whose carbonyl group will develop strong cationic properties. The use of a potassium salt of N-nitroso-N-phenylglycine will delay the formation of the precipitate. Secondly, a nucleophilic attack by the nitrous oxygen on the carbonyl group of the acid will lead to the closure of the (II) ring. Thirdly, an acetate group is lost and a double bond between the two nitrogen atoms is formed (III). Finally, proton loss and the formation of enolic oxygen will produce the final product (IV) (sidnone) (BAKER *et al.*, 1950).

However, Eade and Earl found that the preparation of some analogues of sidnones, such as sidnone containing the nitro group, took a long reaction time and had low yields. They claimed that heating accelerated the formation of the sidnone ring, although it reduced the yield due to the rapid hydrolysis of the product by the hot acidic reaction medium (EADE AND EARL,1946). Therefore, Baker and his colleagues reported for the first time an instantaneous and complete separation of N-arylsidnone with 90% yield when they used trifluoroacetic anhydride (TFAA) as the dehydrating agent (BAKER *et al.*,1950).

The latter synthetic route was successfully used to prepare some complicated sidnones, such as bis-sidnone N with a yield of 70-80%.

Many alternative reagents have also been used to prepare the sidnone system. In 1950, Baker *et al.* used thionyl chloride and reported that the conversion of N-nitroso amino acids to sidnones occurred within a few minutes using thionyl chloride in dry ether at room temperature giving a low yield of 28%. On the other hand, the use of thionyl chloride in a mixture of cold dioxane and pyridine resulted in an improved yield (75%) in 25 min (BAKER *et al.*,1950).

There is a current interest in fluorinated aromatic and heteroaromatic compounds and how this generates a new approach to the synthesis of fluorine-containing sidnones.

Preparation can begin with the corresponding fluorinated pyruvic ester by a step-by-step procedure, as shown in (Figure 17) (ALBOTA *et al.*,2017).

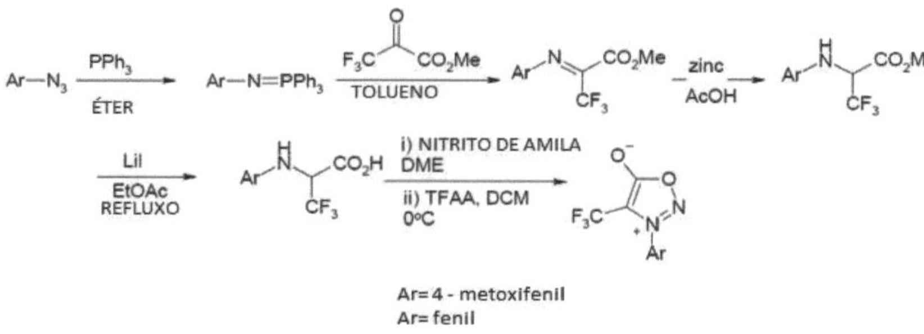

Figure 17. Preparation of sidnones using fluorinated pyruvic ester. Adapted: ALBOTA *et al*,(2017)

5.2.2.4 Sydney and cancer

Mesoionic compounds are known for their possible therapeutic actions and biological activities. These characteristics are related to their structures, represented by betaines, their molecular size and variations in charge density in the ring that extends the exocyclic side chain (ATHAYDE, 2000).

The sidnones (1,2,3-oxadiazol-5-olates) stand out among the class of mesoionic compounds. For more than a decade, researchers have been studying their pharmacological and, in particular, antitumor activities. In addition, sidnones are of great importance because they also act as precursors of bioactive molecules (SHIH *et al.*, 2007).

Although this paper focuses on a review of the antitumor effects of sidnones over the last 10 years, it is important to note that GRYNBERG *et al.* (1992) had already reported the successful use of some sidnone derivatives as antitumor agents *in vivo*. They observed that 3-(4-chloro-3-nitrophenyl) sidnone (SYD-1) and 3-(4-pyrrolidino-3-nitrophenyl) sidnone (SYD-2) increased the survival of mice inoculated with sarcoma 180, Ehrlich carcinoma and fibrous histiocytoma (B10MCII). SYD-2 also showed significant activity against L1210 leukemia. *In in vitro* tests, SYD-1 was the most cytotoxic against all the tumor cells mentioned (GRYNBERG *et al.*, 1992). The mechanisms involved in SYD-1's antitumor activity were not determined by GRYNBERG *et al.* (1992). However, this effect could be related to alterations in mitochondrial energy metabolism, since SYD-1 caused a dysfunction in rat liver mitochondria (HALILA *et al.*, 2007).

GALUPPO *et al. (2016)* evaluated the antitumor activity of SYD-1 (Figure 18) in mice inoculated with the Walker-256 tumor. This carcinosarcoma was discovered by Walker in 1928 from a breast adenocarcinoma in a rat. This tumor can be maintained in the laboratory by successive inoculations. It is one of the few tumor models available for experimental studies in the medical and biological fields in different lines of research (MORAES *et al.*, 2000).

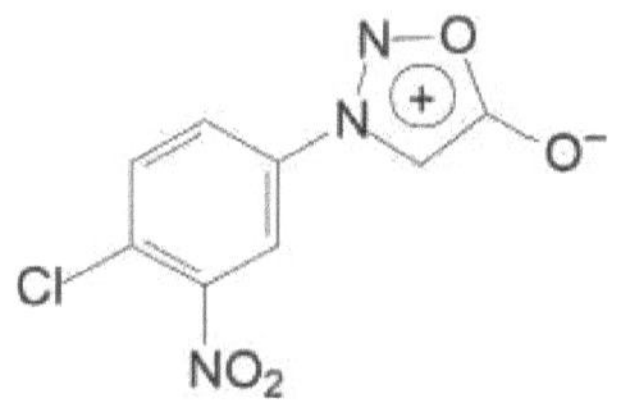

Figure 18. Structure of SYD-1 [N-(4-chloro-3-nitrophenyl)-sidnone]
Source: (GALUPPO et al, 2016).

In this study, tumor cells were inoculated subcutaneously into male Wistar rats. The animals were treated with 75 mg/kg of SYD-1 for 12 days. The volume of the tumor was measured daily after the sixth day, when it became palpable. As a result, considerable decreases in tumor volume and tumor weight were observed in the animals that received SYD-1 (Figures 19 and 20). The group treated with SYD-1 showed a 54% reduction in tumor volume and a 41% reduction in tumor weight. No significant difference was noted in the body weight of the animals in the two groups.

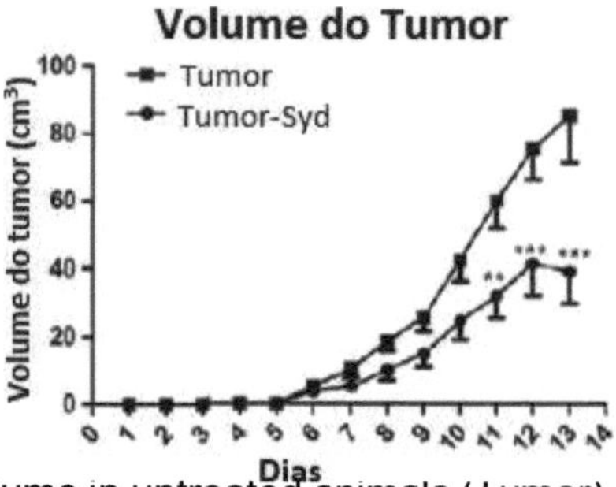

Figure 19. Tumor volume in untreated animals (Tumor) and animals treated with SYD-1 (Tumor-Syd) orally for 12 days. Adapted: GALUPPO *et al.*, (2016)

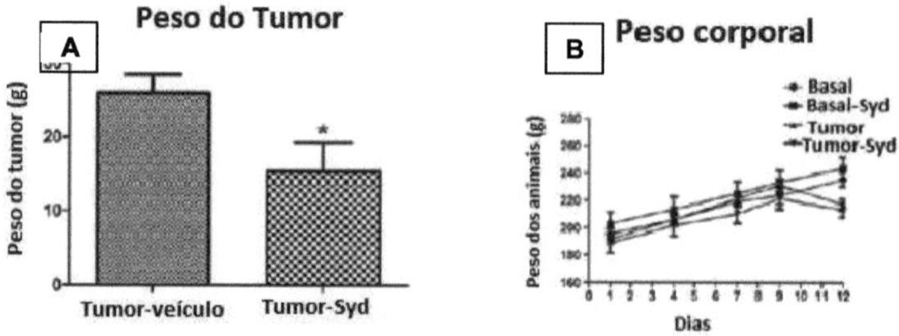

Figure 20. (A) Tumor weight in untreated (Tumor - Vehicle) and SYD-1-treated (Tumor - Syd) animals. (B) Body weight of healthy (Basal), tumor-inoculated (Tumor), healthy SYD-1-treated (Basal-Syd) and tumor-inoculated SYD-1-treated (Tumor-Syd) animals.

Adapted from GALUPPO *et al.* (2016).

Among the biochemical parameters analyzed in the work by GALUPPO et al (2016), we highlight reduced glutathione (GSH), which had its levels decreased in the tumor tissue treated with SYD-1 (Figure 21). GSH is an antioxidant agent and plays an important role in a multitude of cellular processes, including cell differentiation, proliferation, apoptosis and redox homeostasis. Disturbances in GSH homeostasis are involved in the etiology and progression of many human diseases, including cancer (BADRAOUI *et al.* 2009, TRAVESSO *et al*, 2013). The decrease in GSH levels observed by GALUPPO *et al.* (2016) in tumors treated with SYD-1 may represent an important mechanism of action of this molecule in tumor suppression, since the antioxidant capacity of tumor cells may favor tumor progression.

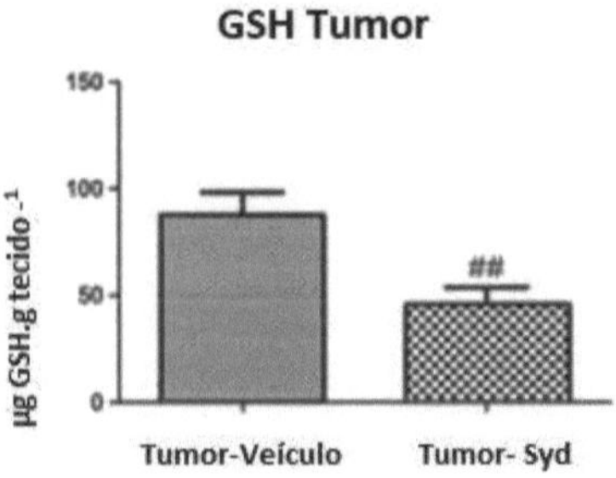

Figure 21. GSH levels in the tumors of untreated (Tumor - Vehicle) and SYD-1-treated (Tumor - Syd) rats. Adapted from GALUPPO *et al.* (2016).

The data from the work by GALUPPO *et al.* (2016) also showed the involvement of the activation of apoptotic pathways and not necrosis in the mechanism of action of SYD-1 (Figure 22). An increase in apoptotic bodies and the expression of pro-apoptotic proteins (Bax and p53) and a reduction in the expression of the anti-apoptotic protein Bcl-2 were observed. Although the results were satisfactory, splenomegaly was observed in the treated animals, which was associated with extravascular hemolysis induced by SYD-1.

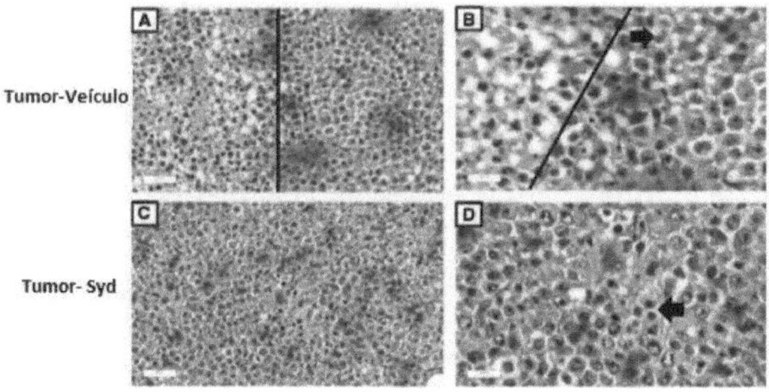

Figure 22. Histological parameters evaluated in rat tumor tissue after 12 days of treatment with SYD-1. Tumor-Veicle (A, B) and Tumor-Syd (C, D). Panel B shows the presence of viable tumor tissue, indicated by the black arrow, and necrosis, indicated by the orange arrow. In the group that received SYD-1 (panel D), the presence of apoptotic bodies can be seen, indicated by the black arrow. Adapted from GALUPPO *et* al. *(*2016).

When it comes to tumor cells, it is essential to talk about the process of apoptosis. This process causes the cell to die and, generally, tumor cells lose control of this mechanism and do not die (WEBER *et al.*, 2004). It is therefore important to find a compound capable of causing apoptosis in tumor cells and thus leading to their death. HOSSAIN *et al.* (2021) evaluated the induction of apoptosis by novel sidnone derivatives. This is the most recent work published on the application of sidnones to tumor cells. The study reported the evaluation of four sidnone derivatives (MC-182, MC-454, MC-433 and MC-431) as possible anticancer agents (Table 1):

Table 1. Siddnone derivatives used, their structures and nomenclatures.

NAME OF COMPOSITE	STRUCTURE	NOMENCLATURE
MC-182	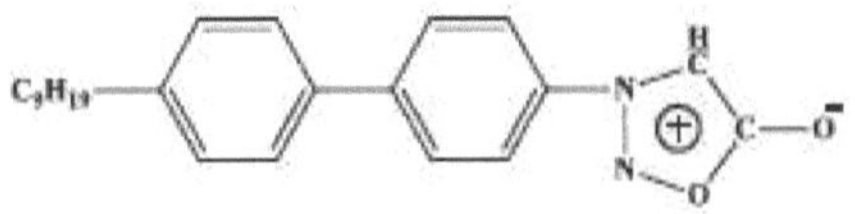	3-(4- nonyl biphenyl- 4' il) sidnone
MC-454	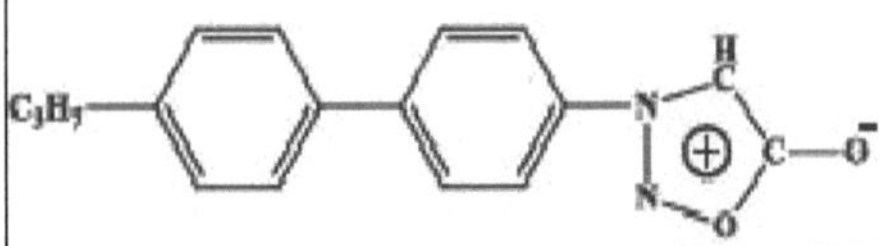	3-(4- propyl biphenyl - 4' il) sidnone
MC-433	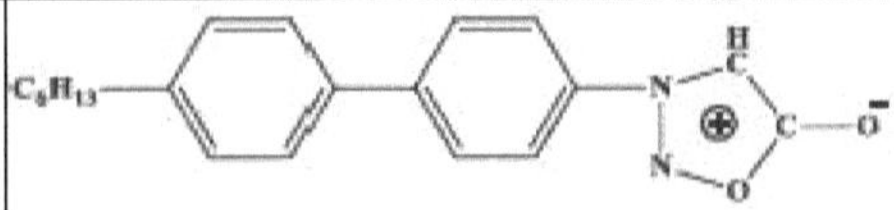	3-(4-hexyl biphenyl-4'-yl) sidnone
MC-431	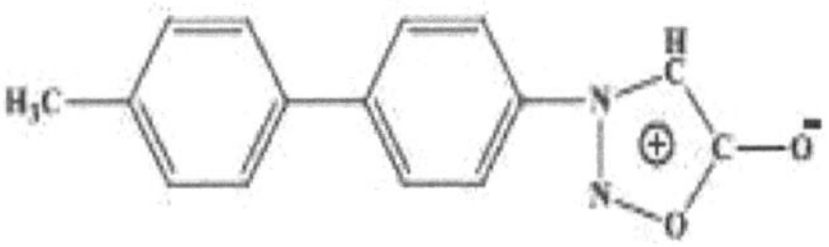	3-(4-methyl biphenyl-4'-yl) sidnone

Adapted from HOSSAIN *et al.*, (2021)

The four compounds were added to breast cancer cell lines (BT 474), human cervical carcinoma cell lines (HeLa) and Jurkat cell lines (human lymphocyte leukemia cells) and the effect on the viability of these cells was evaluated using the MTT ((3-(4,5- dimethyl-2-thiazolyl)-2,5-diphenyltetrazoliobromide) assay.

The compounds showed different effects for each cell line analyzed, as can be seen from their respective IC50 values (concentration of the compound capable of inhibiting 50% of cell viability) (Table 2).

Table 2. IC 50 values for siddnone derivatives in different cell types.

NAME OF COMPOSITE	IC50 (μm) in cells HELA	IC50 (μm) in cells BT 474	IC50 (μm) in cells Jurkat
MC-182	18.19	33.3	42.7
MC-454	46.45	44.83	2.87
MC-433	10.09	33.3	19.08
MC-431	119	1.71	2.73

Adapted from HOSSAIN *et al.,* (2021).

As can be seen in the results of HOSSAIN *et al.* (2021), compound MC-431 showed remarkable antitumor activity against the BT 474 cell line with an IC50 value of 1.71 μM. MC-433, on the other hand, was found to be more effective on the HeLa cell line with an IC50 value of 10.09 μM. MC-454 showed good antitumor activity with an IC50 value of 2.87 μM for the Jurkat cell line. The effect of these compounds on cell viability can be explained by the heterocyclic structure of the compounds used, which have different regions of positive and negative charges. These charges are connected to a polyheteroatomic system, which causes the sidnone molecule to cross cell membranes and have a strong interaction with DNA or proteins. A similar effect was reported in the work by GALUPPO *et al.* (2016) discussed above, in which SYD-1 was tested against Walker-256 carcinosarcoma.

The induction of apoptosis by sidnone derivatives was also evaluated by HOSSAIN *et al.* (2021), incubating HeLa and BT 474 cells with the 4 compounds. This assay makes it possible to distinguish whether cell death is due to necrosis or activation of apoptotic pathways. Apoptotic cell death includes cell morphological changes such as membrane blebbing, formation of apoptotic bodies, contracted nuclei different from the normal cell nucleus, which provide evidence of the apoptotic activity of the compounds (NGUYEN *et al.*, 2014). DAPI (4',6-diamidino-2-phenylindol), a fluorescent DNA-binding agent, annexin V and PI staining techniques were used to make this identification. The analysis was carried out using a microscope to identify death by apoptosis and necrosis qualitatively and quantitatively (Figures 23 and 24).

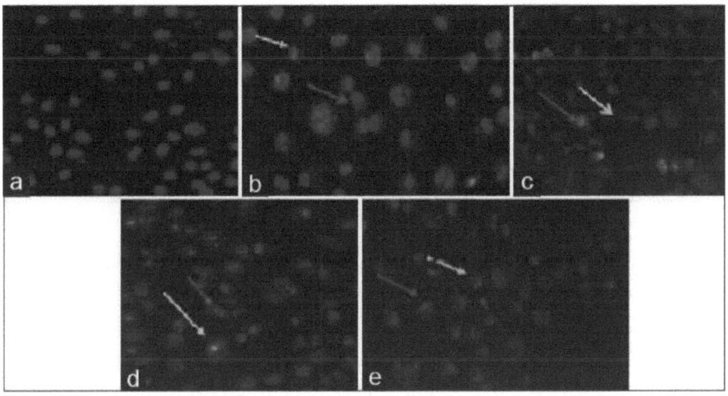

Figure 23. Images seen under a fluorescence microscope. Apoptosis and necrosis in BT-474 cells. (a) BT 474 control cell with DAPI staining; (b) MC-182, (c) MC-454, (d) MC-433, (e) MC-431. Red arrows indicate apoptotic cells and yellow arrows indicate necrotic cells. Source: HOSSAIN *et al.* (2021).

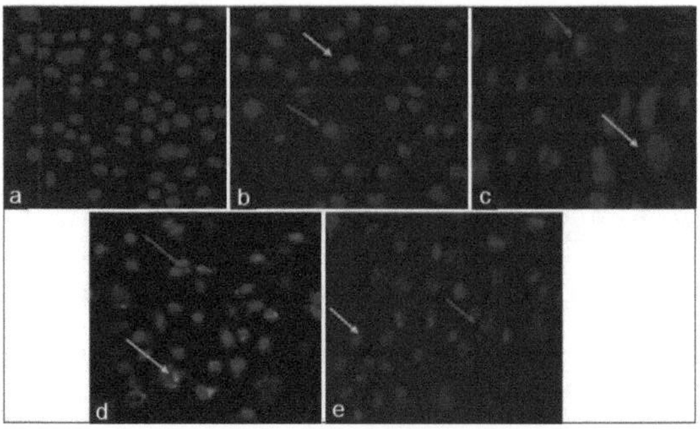

Figure 24. Images seen under a fluorescence microscope. Apoptosis and necrosis in HeLa cells. (a) Control HeLa cell with DAPI staining; (b) MC-182, (c) MC-454, (d) MC-433, (e) MC-431. Red arrows indicate apoptotic cells and yellow arrows indicate necrotic cells. Source: HOSSAIN *et* al., (2021).

Cells with a greenish color are necrotic. Cells stained pink are considered apoptotic cells. Live cells will have no color (Annexin V-/PI-), apoptotic and late apoptotic cells (Annexin V + / PI-) are pink and necrotic cells (Annexin V- / PI +) are green. The results are shown in Tables 3 and 4:

Table 3. Percentage of apoptotic and necrotic cells determined for BT 474 cell lines.

BT 474	CONTROL	MC-182	MC-454	MC-433	MC-431
% cells alive	91.22	47.54	45.3	45.11	46.1
% cells apoptics	2.15	48.32	48.54	48.25	44.34
% cells necrotic	6.63	4.14	6.16	6.64	9.56

Adapted from HOSSAIN *et al.,(*2021).

Table 4. Percentage of apoptotic and necrotic cells determined for Hela cell lines.

HeLa	CONTROL	MC-182	MC-454	MC-433	MC-431
% cells alive	90	48.43	49.23	42.2	48.54
%cells apoptics	3.15	49.15	45.1	14.36	44.32
%cells necrotic	5.49	2.42	5.67	43.44	7.44

The results indicated that all the compounds showed maximum apoptotic and late apoptotic activity in the Hela and BT 474 cell lines, as shown in Figure 23 and 24, except for MC-433 which showed 43% of necrotic cells. TASNEEM *et al.* (2011) used sidnone derivatives to test for anti-tuberculosis activity, antioxidant activity and the main one that will be discussed here, anti-cancer activity. One of the main drugs used in chemotherapy is doxorubicin (trade name: Adriamycin) (HARMAN,1995). However, this drug, like other chemotherapy drugs, has a number of side effects and is ineffective in its selectivity against malignant cells.

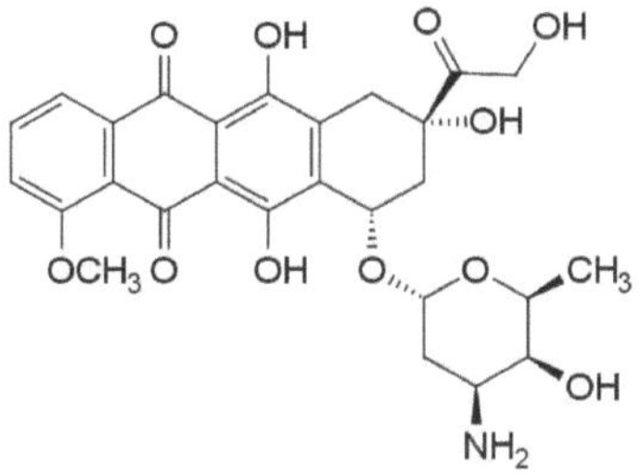

Figure 25. Structure of doxorubicin.

Source: http://www.chem.ucla.edu/~harding/IGOC/D/doxorubicin.html

Over the years, many studies have focused on finding analogues of carbazoles that can act as anticancer agents. The structure of carbazoles has a fused ring and the way they interact with DNA makes them interesting for this function. In this work, the researchers obtained substituted carbazoles by transforming the sidnone ring, which acts as a masked hydrazine. From the Fischer-Indol synthesis, the compounds called 4a-K were prepared (Figure 26).

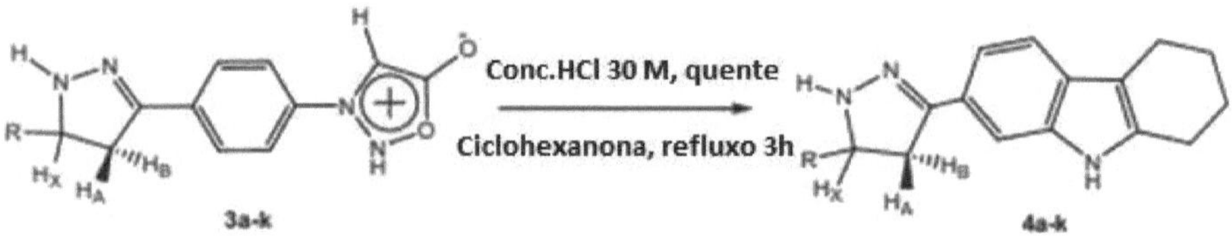

Figure 26. Synthesis of compound 4a-k via 3-[4-(5-aryl-4,5- dihydro-1H-pyrazol-3-yl)]-phenylsidnones (3a-k). Adapted from: TASNEEM *et al.*, (2011).

Compounds 4a, 4b, 4c, 4d, 4e, 4f, 4g, 4h, 4i, 4j and 4k were obtained by changing the R substituent as shown in Table 5. The structures of these compounds were confirmed by IR, ^{1}H NMR, 13C NMR.

Table 5 Substituents of compound 4a-k.

4a	Phenyl
4b	o-chlorophenyl
4c	m-chlorophenyl
4d	p-chlorophenyl
4r	p-nitrophenyl
4f	o-hydroxyphenyl

4g	p-hydroxyphenyl
4h	stylus
4i	methyl
4j	p-anisyl
4k	p-tolila

Adapted from TASNEEM *et al.,* (2011).

In this work, two tumor cell lines were used. One from human kidney cancer (A498) and one from human lung cancer (A549). The results of the cell viability test obtained for each compound against A498 cells can be seen in Tables 6 and 7:

Table 6. Results of the compounds against A498 cells. Percentage (µg/ml) and with ADR as reference. The highlighted values represent greater anticancer activity than the others.

COMPOSITE	10	20	40	80
4a	100	99.1	95.9	3.9
4b	100	100	97.4	16.2
4c	100	100	97.9	33.5
4d	100	100	97.5	31.7
4e	100	100	97.7	4.1
4f	99.3	99.5	97.7	16.5
4g	83.0	86.0	83.0	3.9
4h	99.5	99.7	95.7	6

4i	80.0	82.0	85.0	4
4j	98.3	100	97.7	4.2
4k	99.3	99.2	96.2	4.6
ADR	10.7	8.2	-39.5	-66.9

Adapted from TASNEEM et al (2011).

Table 7. Results of the compounds against A549 cells. Percentage (μg/ml) and with ADR as reference. The highlighted values represent greater anticancer activity than the others.

COMPOSITE	10	20	40	80
4[a]	98.8	94.8	88.4	48.7
4b	90.2	81.6	75.7	12.1
4c	81.6	76.2	69.4	15.3
4d	83.2	74.6	67.1	11.5
4e	70.1	64.4	50.4	4.2
4f	11.9	-2.9	-18.5	-28.4
4g	83.2	74.6	67.1	11.5
4h	98.8	94.8	88.4	48.7
4i	90.2	81.6	75.7	12.1
4j	81.6	76.2	69.4	15.3
4k	11.9	-2.9	-18.5	-28.4
ADR	10.7	8.2	-39.5	-66.9

Adapted from TASNEEM et al (2011).

Using adriamycin (ADR) as a reference, few compounds showed antitumor activity. Compounds 4g and 4i, with a hydroxyl group in the para position of the ring and a methyl group, showed partial activity in the A498 cell line, while compounds 4f and 4g, with hydroxyl in the ortho and para positions of the ring, showed partial activity in A549 cells.Figures 27 and 28 represent the results obtained:

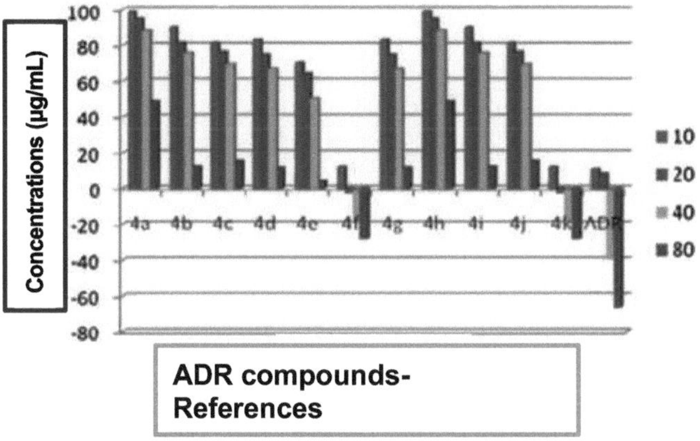

Figure 27. Anticancer activity of the compounds synthesized 4a-k on the A498 cell line. Adapted from: TASNEEM *et al.,* (2011).

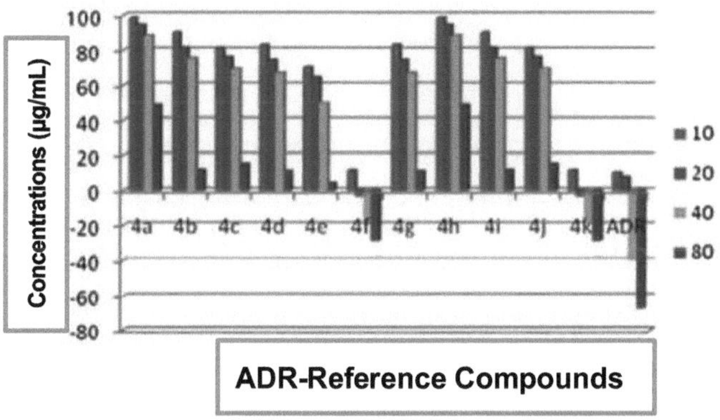

Figure 28. Anticancer activity of the compounds synthesized 4a-k in the A549 cell line. Adapted from: TASNEEM *et al.,* (2011).

The sidnones were also studied as starting material in the creation of the synthetic route for the production of pyrazole-based analogues of

Combretastatin A4 (CA4P) (Figure 29) (BROWN *et al.*, 2016). Combretastatins are a group of compounds isolated by Pettit (1982) from the bark and stem of the tree belonging to the *Combretaceae* family, native to riverbanks in the Eastern Cape Province (PETIT, 1982).

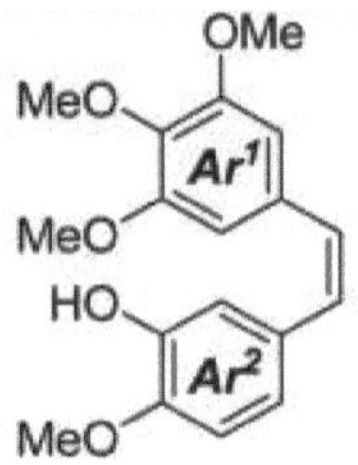

Figure 29. Structure of Combrestina A4. Source: BROWN *et al.,* (2016).

CA4P is classified as an anti-angiogenic agent because it develops the ability to rapidly interact in a specific way with the abnormal vascularization of tumours, resulting in vascular collapse and tumour necrosis. This precise activity mediated through cadherin-vascular endothelium signaling acts through two mechanisms, the first preventing the formation of new vessels that will feed the tumor and the second, acting on vessels already formed (FRAGA *et al.*, 2015).

Pyrazole is an important fragment in the pharmaceutical industry and is present in a number of commercial drugs and pesticides (SINGH *et al.*, 2009). However, synthetic routes for this compound suffer from regioselectivity problems during the heterocycle formation stage. For this reason, the researchers assessed that sidnones have the potential to be starting compounds for these desirable therapeutic analogues. Sidnones undergo [3 + 2] cycloaddition and these can be highly regioselective. The authors proposed routes for the formation of pyrazole analogues (Figure 30):

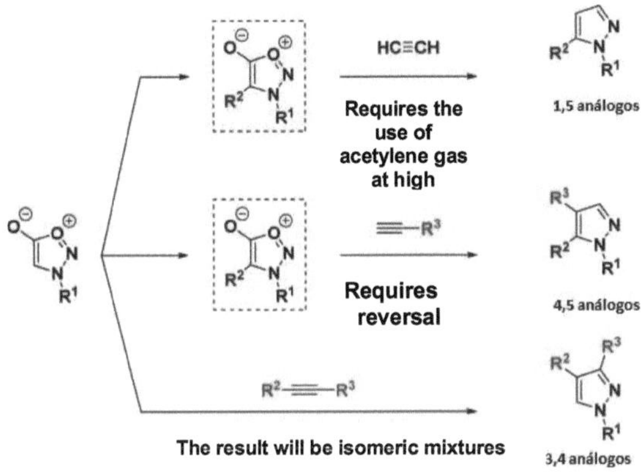

Figure 30. Proposed routes for the formation of CA4 analogues. Adapted from BROWN *et al.,* (2016).

The syntheses were carried out and through quantitative and morphological results, they arrived at compound 31, which showed the highest yield (Figure 31).

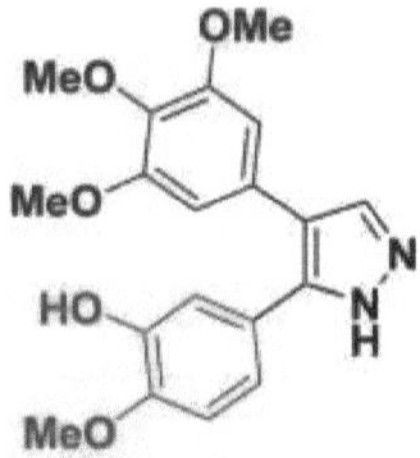

31

Figure 31. Compound named 31. Source: BROWN *et al.,* (2016).

For the tests on antitumor efficiency, human rectal colon adenocarcinoma (SW1222) was inoculated into the back of female mice. When the tumor reached 8 mm in diameter, the treatments were started.

This group was divided into 3 subgroups: the first, (control), received 50% Na2CO3 /NaCl, 10 mL / kg. The second group received CA4P (100 mg / kg, 0.227 mmol / kg, in 50% Na2CO3 / NaCl, 10 mL / kg and the third group received compound 31 (81 mg / kg, 0.227 mmol / kg, 50% Na2CO3 / NaCl, 10 mL / kg). All treatments were administered as a single dose.

The results on the effects of the treatments on the tumor cells were quantified after 24 hours (Figure 32). Compound 31 showed a tendency to increase necrosis in a similar way to CA4P at an equivalent dose, without the need for solubilization in water.

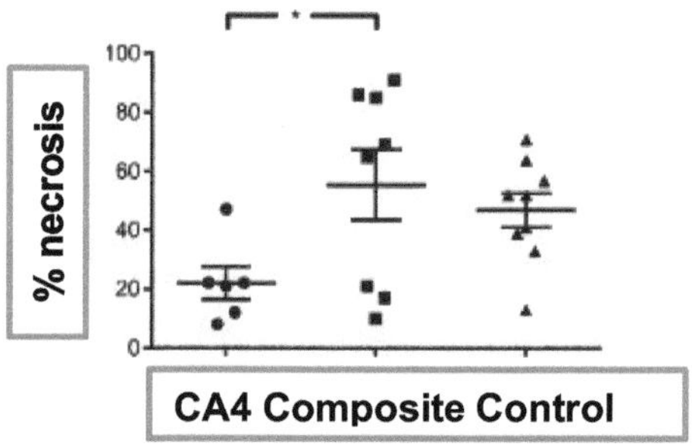

Figure 32. Result of tumor treatment comparing CA4P and compound 31. Each point represents the combined data from a single tumor. Adapted: BROWN *et al.,* (2016).

Sidnones also have wide applicability as NO donors. Nortcliffe *et al.* (2014) synthesized a sequence of peptides conjugated to NO donor groups and evaluated their anticancer activity in comparison to abiraterone, which is a drug for the treatment of prostate cancer. They obtained positive cytotoxicity results against prostate cancer and breast cancer cell lines. This role of sidnone as an NO donor was also described by Pepelová *et al.* (2016), where Molsidomine, an NO-releasing drug considered to be cardioprotective, was metabolized by morpholinosidnonimine in cardiotoxicity tests. Both studies do not use the sidnones themselves for treatment, but take advantage of their structures to obtain active compounds that will act in tests.

6 CONCLUSION

Sidnones are compounds with special structures that give them particular characteristics, thus providing various pharmacological activities. Among these activities, anticancer stands out.

Cancer is one of the leading causes of death worldwide and the treatments currently available are not entirely effective, but have a number of side effects, including toxicity to normal cells.

The articles discussed show promising results for the use of sidnones as tumor agents. Although the results show a potential for the use of these molecules in cancer therapy, more studies are needed on the mechanisms of action of these compounds and their safety for healthy cells.

REFERENCES

Abdualkader, Abdualrahman & Taher, Muhammad & Yusoff, Nik. (2017). Mesoionic sydnone: a review in their chemical and biological properties. International Journal of Pharmacy and Pharmaceutical Sciences. Albota, Florin & Stănescu, Michaela (2017). The state of art in sydnones chemistry and applications. Revue Roumaine de Chimie. 62. 711-734.

Applegate J, Turnbull K. The efficient synthesis of 3-arylsydnones under neutral conditions. Synthesis 1988;1988:1011-2.

BADAMI, B. V. Mesoionic Compounds: An Unconventional Class of Aromatic Heterocycles. *Resonance,* vol. 11, *Indian Academy of Sciences*, p. 40-48. 2006.

Badraoui R, Blouin S, Moreau MF, Gallois Y, Rebai T, SahnounZ et al. Effect of alpha tocopherol acetate in Walker 256/B cells-induced oxidative damage in a rat model of breast cancer skeletalmetastases. Chem Biol Interact 2009;182:98- 105

Baker W, Ollis WD, Poole VD. Cyclic mesoionic compounds. Part III. Further properties of the sydnones and the mechanism of their formation. J Chem Soc (Resumed); 1950. p. 1542-51.

BRANDT, A. P.; GOZZI, G. J.; PIRES, A. R. A.; MARTINEZ, G. R.; CANUTO, A . V. S.; ECHEVARRIA, A.; PIETRO, A. D. P.; CADENA, S. M. S. C. Impairment of Oxidative Phosphorylation Increases the Toxicity of SYD-1 on Hepatocarcinoma Cells (HepG2). Chemico-Biological Interactions, vol. 256, p. 154-160, 2016.

BRAY, F. et al. Global cancer statistics 2018: GLOBOCAN estimates of incidence and mortality worldwide for 36 cancers in 185 countries. CA: a cancer journal for clinicians, Hoboken, v. 68, n. 6, p. 394-424, Nov. 2018

BRAZ, V. Synthesis, Structure and Kinetics of SNAr Reactions of Aryl-Sidnones. Thesis (Master's Degree in Chemistry, Exact Sciences). Chemistry Institute, Federal Rural University of Rio de Janeiro, Seropédica - RJ. 1992.

Brown AW, Fisher M, Tozer GM, Kanthou C, Harrity JP. Sydnone

Cycloaddition Route to Pyrazole-Based Analogs of Combretastatin A4. J Med Chem. 2016 Oct 27;59(20):9473-9488. doi: 10.1021/acs.jmedchem.6b01128. Epub 2016 Oct 7.

Chandrasekhar R, Nanjan MJ. Sydnones: a brief review. Mini Rev Med Chem. 2012 Nov;12(13):1359-65. doi: 10.2174/13895575112091359.

Chem Biol Interact 2007;160:160-70 Badraoui R, Blouin S, Moreau MF, Gallois Y, Rebai T, Sahnoun.

Chem Biol Interact 2007;160:160-70 Halila CG, Oliveira MBM, Echevarria A, Bel em AC, Rocha.

D. Grossie, L. Harrison and K. Turnbull, *Acta Cryst.*, 2013, *E69*, o1196; D. Grossie, L. Harrison and K. Turnbull, *Acta Cryst.*, 2014, *E70*, o1165-o1166.

Eade RA, Earl JC. The sydnones. a new class of compound containing two adjacent nitrogen atoms. J Chem Soc (Resumed); 1946. p. 591-93.

Earle WR. A study of the Walker rat mammary carcinoma 256, in vivo and in vitro. Am J Cancer 1935;24:566-612.

Galuppo LF, Dos Reis Lívero FA, Martins GG, Cardoso CC, Beltrame OC, Klassen LM, Canuto AV, Echevarria A, Telles JE, Klassen G, Acco A. Sydnone 1: A Mesoionic Compound with Antitumoral and Haematological Effects In Vivo. Basic Clin Pharmacol Toxicol. 2016 Jul;119(1):41-50. doi: 10.1111/bcpt.12545. Epub 2016 Jan 25.

GOZZI, G. J. J.; PIRES, A. R.; VALDAMARI, G.; ROCHA, E. M.; MARTINEZ, G. R.; NOLETO, G. R.; ACCO, A.; SOUZA, C. E. A.; ECHEVARRIA, A.; REIS, C. M.DIPIETRO, A.; CADENA, S. M. S. C. Selective cytotoxicity of mesoionic 1,3,4-thiadiazolium derivatives in hepatocarcinoma cells (HepG2). *Plos one,* p. 1-17. 2015

Grynberg N, Gomes R, Shinzato T, Echevarria A, Miller J. Some new aryl-sydnones: effects on murine tumors. Anticancer Res. 1992 May-Jun;12(3):1025- 8.

Halila CG, Oliveira MBM, Echevarria A, Belem AC, Rocha MEM, Carnieri EGS et al. Effect of sydnone SYD-1, a mesoioniccompound, on energy-linked functions of rat liver mitochondria.Chem Biol Interact 2007;160:160-

70

Halila CG, Oliveira MBM, Echevarria A, Bel em AC, Rocha MEM, Carnieri EGS et al. Effect of sydnone SYD-1, a mesoionic compound, on energy-linked functions of rat liver mitochondria.

Hossain SL, Mathews M, Bhyranalyar Nagarajappa VS, Kumar BK, Veerappa Yelamaggad CV, Singh CR. Antiproliferative, apoptosis-inducing activity and molecular docking studies of sydnones compounds. J Cancer Res Ther. 2021 Oct 25.

INCA. Cancer statistics. Available at: <https://www.inca.gov.br/numeros-de-cancer>. Accessed on: October 28, 2021.

INCA. What is cancer? Available at: <https://www.inca.gov.br/o-que-e-cancer>. Accessed on: October 28, 2021.

INCA. Types of cancer: Available at: <https://www.inca.gov.br/tipos-de-cancer>. Accessed on October 28, 2021.

KAI-CHUNG, K.; ECHEVARRIA, A.; GALLEMBECK, S.; MACIEL, M.A.; MILLER, J.; RUMJANEK, V.; SIMAS, A. Mesoionic compounds 4. Structure of 1,4,5-triphenyl- 1,2,4-triazolium-3-thiolate. *Acta Cryst.*, v.49, p.1092-1094, 1993.

Kenner J, Mackay K. Structure of the sydnones. Nature 1946;21:909-10.

Kier LB, Roche E. Molecular orbital calculations of the electronic structure of the sydnones. J Pharm Sci 1966;55:807-12.

Lenčová-Popelová O, Jansová H, Jirkovský E, Bureš J, Jirkovská-Vávrová A, Mazurová Y, Reimerová P, Vostatková L, Adamcová M, Hroch M, Pokorná Z, Kovaříková P, Šimůnek T, Štěrba M. Are cardioprotective effects of NO-releasing drug molsidomine translatable to chronic anthracycline cardiotoxicity settings? Toxicology. 2016 Nov 30;372:52-63.

Lira, B. F.; Athayde Filho, P. F.; Miller, J.; Simas, A. M.; Dias, A. F.; Vieira, M. J. Synthesis and characterization of some new mesoionic 1,3-thiazolium-5- thiolates via cyclodehydration and *in situ* 1,3-dipolar cycloaddition/cycloreversion. *Molecules*, vol. 7, p. 791-800. 2002

MEM, Carnieri EGS et al. Effect of sydnone SYD-1, a mesoionic compound, on energy-linked functions of rat liver mitochondria.

Moraes SP, Cunha A, Reis Neto JA, Barbosa H, Roncolatto CAP, Duarte RF. Experimental model of Walker's tumor. Acta Cir Bras [serial online] 2000 Oct- Dec;15(4).

NEWTON, C. G., RAMSDEN, C.A. Mesoionic heterocycles. *Tetrahedron*, v.38, n.20, p.2965-3011, 1982.

Nguyen B, Hamelberg D, Bailly C, Colson P, Stanek J, Brun R, *et al.* Characterization of a novel DNA minor-groove complex. Biophys J 2004;86:1028-41

Nortcliffe, A, Fleming, IN, Botting, NP & O'Hagan, D 2014, 'Synthesis and anticancer properties of RGD peptides conjugated to nitric oxide releasing functional groups and abiraterone', *Tetrahedron*, vol. 70, no. 44, pp. 8343-8347.

OLLIS, W.D.; RAMSDEN, C.A. Mesoionic compounds. Advances in Heterocyclic Chemistry, v.19, p. 120-121, 1976.

Patra N, De U, Kang JA, Kim JM, Ahn MY, Lee J, Jung JH, Chung HY, Moon HR, Kim HS. A novel epoxypropoxy flavonoid derivative and topoisomerase II inhibitor, MHY336, induces apoptosis in prostate cancer cells. Eur J Pharmacol.
2011 May 11;658(2-3):98-107.

Pettit, G. R.; Cragg, G. M.; Herald, D. L.; Schmidt, J. M.; Lohavanijaya, P. Isolation and structure of combretastatin. Canadian Journal of Chemistry 1982, 60, 1374.

Pilli HG, Safak C, Abbasoğlu U. Sydnone derivatives: synthesis and antimicrobial activity. Arch Pharm (Weinheim). 1993 Sep;326(9):559-61.

QUEIROZ, W. Avaliação da Ação Citotóxica de Diferentes Séries de Compostos Derivados Mesoiônicos em Células Derivadas de Hepatocarcinoma Humano (HepG2).Monografia (Licenciatura em química, Ciências exatas). Chemistry Institute, Federal Rural University of Rio de Janeiro, Seropédica - RJ. 2020

ROBBINS, S. L.; COTRAN, R. S. Pathologic Basis of Disease. 7th edition. Philadelphia: Elsevier Saunders. 2004.

Shih MH, Chen JC, Lin GL, Lin TT, Sun MH. Novel synthesis of palladium (II) complexes derived from 3-arylsydnone-4-carbaldehyde N-phenylthiosemicarbazones and biological activity. J Pharm Pharmacol 2014;66:73-83.

SILVA, G. R.; SOUZA, B. F.; PIUVEZAM, M. R. Mesoionic compounds and their antitumor activity: Literature Review. Revista Saúde e Ciência Online. 2018.

SOUSA-PEREIRA, D. Synthesis, characterization and anticancer activity of hydrazino-carbothioamides and mesoionic compounds of the 1,3,4-thiadiazolium-2-aminide class. Thesis (Doctorate in Chemistry, Exact Sciences). Institute of Chemistry, Federal Rural University of Rio de Janeiro, Seropédica - RJ. 2019.

Sung H, Ferlay J, Siegel RL, Laversanne M, Soerjomataram I, Jemal A, Bray F. Global Cancer Statistics 2020: GLOBOCAN Estimates of Incidence and Mortality Worldwide for 36 Cancers in 185 Countries. CA Cancer J Clin. 2021 May;71(3):209-249.

Taj T, Kamble RR, Gireesh TM, Hunnur RK, Margankop SB. One-pot synthesis of pyrazoline derivatized carbazoles as antitubercular, anticancer agents, their DNA cleavage and antioxidant activities. Eur J Med Chem. 2011 Sep;46(9):4366- 73.

Z et al. Effect of alpha tocopherol acetate in Walker 256/B cells-induced oxidative damage in a rat model of breast cancer skeletal metastases. Chem Biol Interact 2009;182:98-105

MIX
Papier aus verantwortungsvollen Quellen
Paper from responsible sources
FSC® C105338

Printed by Books on Demand GmbH, Norderstedt / Germany